short tail
bear cub

busy tail
dog

spotted tail
snow leopard

straight tail
meerkat

bushy tail
skunk

long tail
cow

colorful tail
peacock

for Janie and Carol, friends for a lifetime—ST

For Leanne, Emilia, Mum, Dad, Claire, and Kiba.
Thank you for everything—MG

Library of Congress Cataloging-in-Publication data is on file with the publisher.

Text copyright © 2019 by Sue Tarsky
Illustrations copyright © 2019 by Albert Whitman & Company
Illustrations by Michael Garton
First published in the United States of America in 2019 by Albert Whitman & Company
ISBN 978-0-8075-9045-4 (hardcover)
ISBN 978-0-8075-9048-5 (ebook)

Printed in China
10 9 8 7 6 5 4 3 2 1 HH 24 23 22 21 20 19
Design by Aphee Messer

For more information about Albert Whitman & Company,
visit our website at www.albertwhitman.com.

100 Years of Albert Whitman & Company
Celebrate with us in 2019!

Whose Are These?
Whose Tail?

Sue Tarsky  illustrated by **Michael Garton**

Albert Whitman & Company
Chicago, Illinois

long tail

short tail

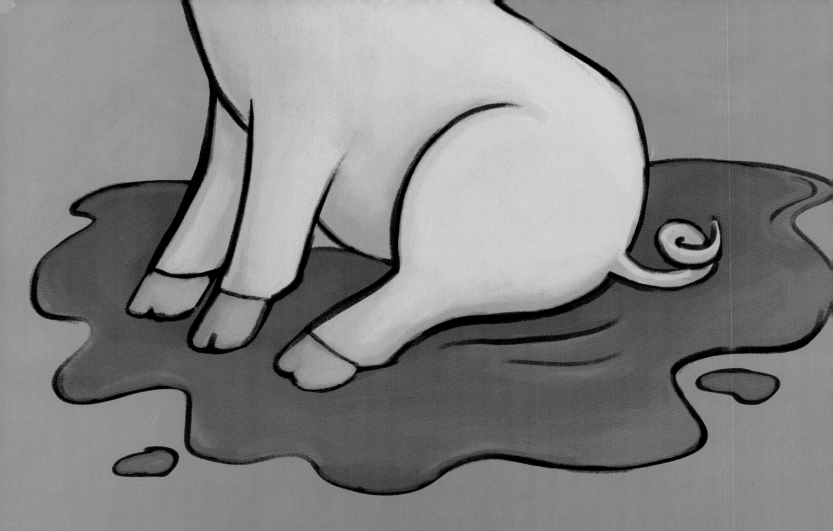

curly tail

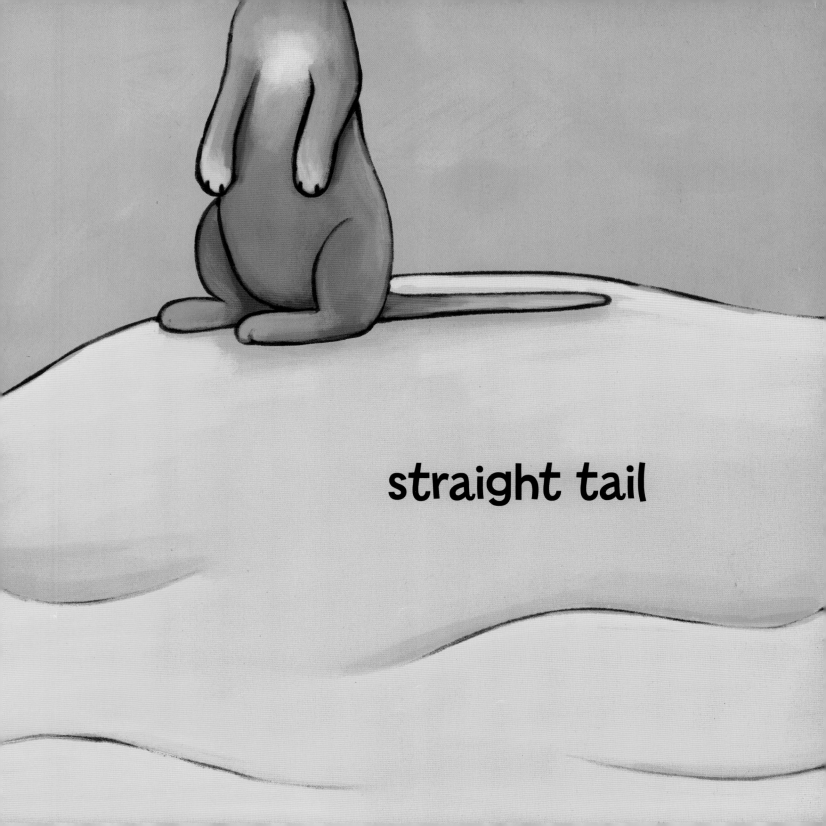

straight tail

busy tail

still tail

swishing tail

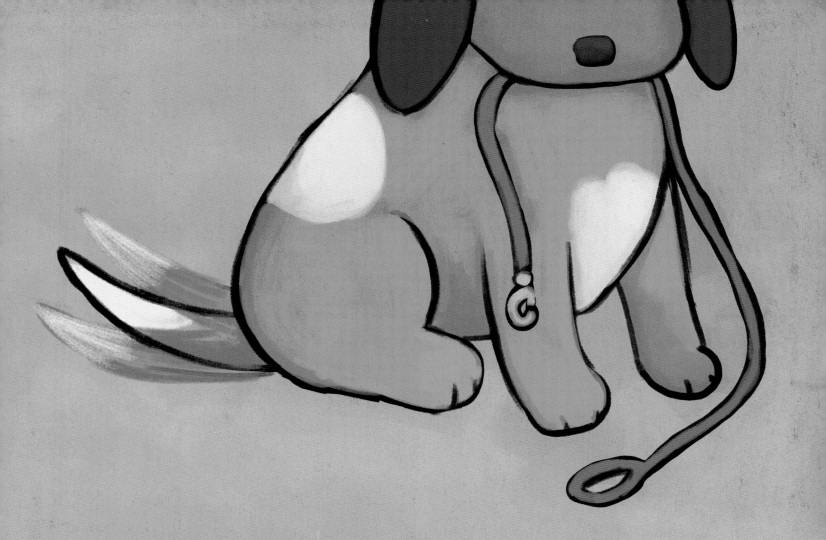

thumping tail

striped tail

spotted tail

bushy tail

fluffy tail

white tail

colorful tail

no tail

my tail!

white tail
sheep

swishing tail
horse

striped tail
raccoon

still tail
cat

thumping tail
puppy

fluffy tail
bunny

curly tail
pig